Everyday Mathematics®

The University of Chicago School Mathematics Project

Student Math Journal
Volume 1

Grade  1

 **Education**

Chicago, IL • Columbus, OH • New York, NY

The University of Chicago School Mathematics Project (UCSMP)

Max Bell, Director, UCSMP Elementary Materials Component; Director, *Everyday Mathematics* First Edition; James McBride, Director, *Everyday Mathematics* Second Edition; Andy Isaacs, Director, *Everyday Mathematics* Third Edition; Amy Dillard, Associate Director, *Everyday Mathematics* Third Edition; Rachel Malpass McCall, Associate Director, *Everyday Mathematics* Common Core State Standards Edition

Authors

Max Bell, Jean Bell, John Bretzlauf, Amy Dillard, Robert Hartfield, Andy Isaacs, James McBride, Rachel Malpass McCall, Kathleen Pitvorec, Peter Saecker

Technical Art
Diana Barrie

Third Edition Teachers in Residence
Jeanine O'Nan Brownell, Andrea Cocke, Brooke A. North

UCSMP Editorial
Rossita Fernando, Lila K. Schwartz

Contributors

Allison Greer, Meg Schleppenbach, Cynthia Annorh, Robert Balfanz, Judith Busse, Mary Ellen Dairyko, Lynn Evans, James Flanders, Dorothy Freedman, Nancy Guile Goodsell, Pam Guastafeste, Nancy Hanvey, Murray Hozinsky, Deborah Arron Leslie, Sue Lindsley, Mariana Mardrus, Carol Montag, Elizabeth Moore, Kate Morrison, William D. Pattison, Joan Pederson, Brenda Penix, June Ploen, Herb Price, Dannette Riehle, Ellen Ryan, Marie Schilling, Susan Sherrill, Patricia Smith, Kimberli Sorg, Robert Strang, Jaronda Strong, Kevin Sweeney, Sally Vongsathorn, Esther Weiss, Francine Williams, Michael Wilson, Izaak Wirzup

Photo Credits

Cover (l)C Squared Studios/Getty Images, (r)Tom & Dee Ann McCarthy/CORBIS, (bkgd)Ralph A. Clevenger/CORBIS; **Back Cover** C Squared Studios/Getty Images; **34** C Squared Studios/Photodisc/Getty Images; **others** The McGraw-Hill Companies.

---

everyday**math**.com

STEM McGraw-Hill is committed to providing instructional materials in Science, Technology, Engineering, and Mathematics (STEM) that give all students a solid foundation, one that prepares them for college and careers in the 21st century.

Send all inquiries to:
McGraw-Hill Education
STEM Learning Solutions Center
P.O. Box 812960
Chicago, IL 60681

ISBN: 978-0-07-657729-0
MHID: 0-07-657727-9

Printed in the United States of America.

5 6 7 8 9 QVR 17 16 15 14 13 12 11

The McGraw-Hill Companies

Contents

UNIT 1 Establishing Routines

UNIT 2 Everyday Uses of Numbers

UNIT 3 Visual Patterns, Number Patterns, and Counting

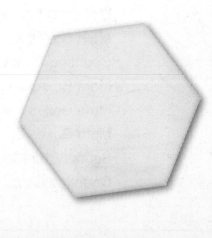

UNIT 4 Measurement and Basic Facts

UNIT 5   Place Value, Number Stories, and Basic Facts

Activity Sheets

Koala
19 lb

Cheetah
120 lb

Penguin
75 lb

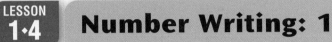

Number Writing: 1

LESSON 1·4

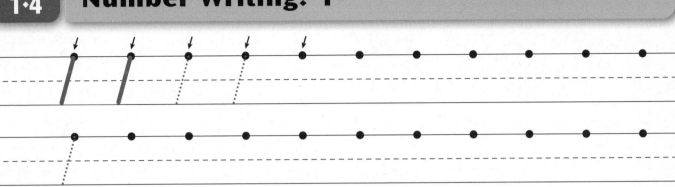

1		
	$1 + 0$	
/	$2 - 1$	
uno		one

Draw a picture of 1 thing.

Number Writing: 2

LESSON 1·4

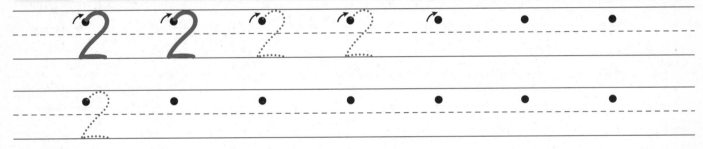

2		
	$1 + 1$	
//	$3 - 1$	
dos		two

Draw a picture of 2 things.

LESSON 1·7 **Number Writing: 3**

33 3 3 3 • • •

3 • • • • • • •

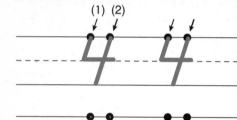

Draw a picture of 3 things.

 LESSON 1·7 **Number Writing: 4**

(1) (2)
4 4 4 4 4 • • • • • •

4 • • • • • • • • • •

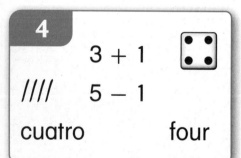

Draw a picture of 4 things.

LESSON 1·8 Dice-Roll and Tally

Roll a die. Use tally marks to record the results on this chart.

	Tallies	Total
⚀	~~IIII~~ ~~IIII~~ I	II II
⚁	~~IIII~~ III	8
⚂	~~IIII~~ III	8
⚃	IIII	4
⚄	~~IIII~~ ~~IIII~~	10
⚅	~~IIII~~ I	6

Calendar

Month _____

Sunday	Monday	Tuesday	Wednesday	Thursday	Friday	Saturday

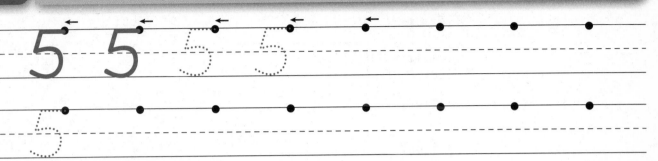

LESSON 1·9 Number Writing: 5

5 5 5 5

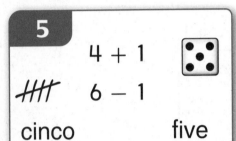

5	
4 + 1	
☰ 6 − 1	
cinco	five

Draw a picture of 5 things.

LESSON 1·9 Number Writing: 6

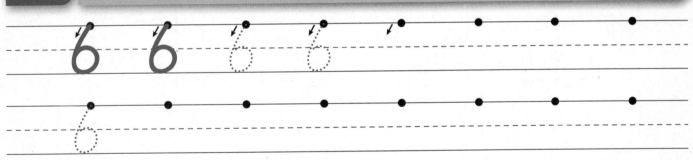

6 6 6 6

6	
5 + 1	
☰ / 7 − 1	
seis	six

Draw a picture of 6 things.

LESSON 1·12 A Thermometer

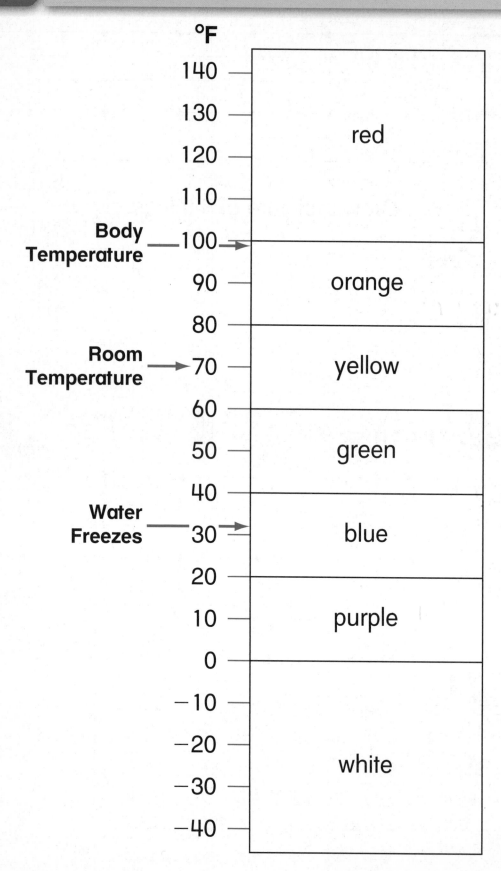

°F

- 140
- 130
- red
- 120
- 110
- Body Temperature → 100
- 90 orange
- 80
- Room Temperature → 70 yellow
- 60
- 50 green
- 40
- Water Freezes → 30 blue
- 20
- 10 purple
- 0
- −10
- −20
- white
- −30
- −40

LESSON 2·1 *Rolling for 50*

Materials

◆ a die

◆ a marker for each player

◆ a gameboard

Players 2

Skill Count by 1s

Object of the Game

To be the first player to reach 50

Directions

Take turns.

1. Put your marker on 0.

2. Roll the die. Look in the table to see how many spaces to move.

3. The first player to reach 50 wins.

Roll	Spaces
1	3 up
2	2 back
3	5 up
4	6 back
5	8 up
6	10 up

| 0 |

| 1 | 2 | 3 | 4 | 5 | 6 | 7 | 8 | 9 | 10 |

| 11 | 12 | 13 | 14 | 15 | 16 | 17 | 18 | 19 | 20 |

| 21 | 22 | 23 | 24 | 25 | 26 | 27 | 28 | 29 | 30 |

| 31 | 32 | 33 | 34 | 35 | 36 | 37 | 38 | 39 | 40 |

| 41 | 42 | 43 | 44 | 45 | 46 | 47 | 48 | 49 | 50 |

LESSON 2·2 Information about Me

My first name is _____.

My second name is _____.

My last name is _____.

I am _____ years old.

Put candles on your cake.

My area code and home telephone number are

(_____ _____ _____) _____ _____ _____ – _____ _____ _____ _____

    (area code)   (telephone number)

Important Phone Numbers

Emergency number: _____ _____ _____

School number:

(_____ _____ _____) _____ _____ _____ – _____ _____ _____ _____

Local library number:

(_____ _____ _____) _____ _____ _____ – _____ _____ _____ _____

LESSON 2·2 Addition Facts Record

8A

_____ + _____ = _____   _____ + _____ = _____

_____ + _____ = _____   _____ + _____ = _____

_____ + _____ = _____   _____ + _____ = _____

_____ + _____ = _____   _____ + _____ = _____

_____ + _____ = _____   _____ + _____ = _____

_____ + _____ = _____   _____ + _____ = _____

Subtraction Facts Record

_____ − _____ = _____   _____ − _____ = _____

_____ − _____ = _____   _____ − _____ = _____

_____ − _____ = _____   _____ − _____ = _____

_____ − _____ = _____   _____ − _____ = _____

_____ − _____ = _____   _____ − _____ = _____

_____ − _____ = _____   _____ − _____ = _____

Date _____

Number Writing: 7

7		
6 + 1		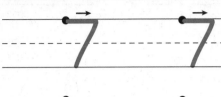
ĦĦ //		8 − 1
siete		seven

Draw a picture of 7 things.

Number Writing: 8

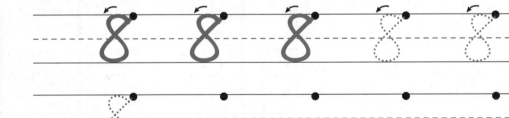

8		
7 + 1		
ĦĦ ///		9 − 1
ocho		eight

Draw a picture of 8 things.

1. Count up by 1s.

    7,   8,   9,

_____, _____, _____,

_____, _____, _____,

_____, _____

2. Count up by 5s.

    0,   5,   10,

_____, _____, _____,

_____, _____, _____

3. Write the number that comes before.

_____ 19

_____ 23

_____ 31

_____ 36

4. Write the number.

_____   _____

Date

Number Writing: 9

9 9

8 + 1

    10 − 1

nueve   nine

Draw a picture of 9 things.

Number Writing: 0

0 0

0 + 0

1 − 1

cero   zero

Date _____

---

1. How many tally marks?

~~HHH~~ ~~HHH~~ //

Choose the best answer.

◯ 3

◯ 12

◯ 15

◯ 10

2. Draw the shape you are more likely to grab from the bag.

_____

3. Make a sum of 10 pennies.

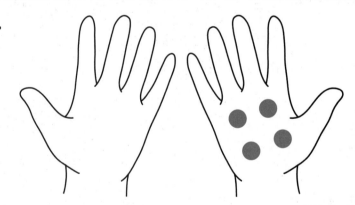

4. Complete the number line.

8   9   ___ ___ ___ ___ ___ ___

---

12   twelve

Date _____

1. Count back by 1s.

  18,   17,   16,

  _____, _____, _____,

  _____, _____, _____,

  _____, _____

2. Count up by 5s.

  10,   15,   20,

  _____, _____, _____,

  _____, _____

3. What number comes before 10?

Choose the best answer.

 1

 0

⬭ 9

⬭ 11

4. Write the number.

_____   _____

LESSON 2·6

Telling Time

1. Record the time.

_____ o'clock

_____ o'clock

_____ o'clock

_____ o'clock

2. Draw the hour hand.

2 o'clock

6 o'clock

14 fourteen

Math Boxes

1. How many tally marks?

HHT HHT HHT ////

_____ tally marks

2. Draw the shape you are more likely to grab from the bag.

_____

3. Make a sum of 10 pennies.

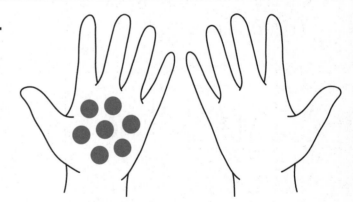

4. Complete the number line.

22   23   ___   ___   ___   ___   ___

Date _____

1. Record the time.

_____ o'clock

2. Use your number grid.

Start at 12.

Count up 5.

You end at _____.

3. Circle the winning card in *Top-It*.

11   9

4. What day of the week is today?

_____

What day of the month?

_____

What day of school?

_____

LESSON 2·8 Math Boxes

1. Use your number grid.

Start at 11.

Count back 6.

You end at _____.

2. How much money?

Ⓟ Ⓟ Ⓟ Ⓟ Ⓟ Ⓟ

_____ ¢

3. What comes next?

△ ○ △ ○ _____

Choose the best answer.

 △

 □

 ○

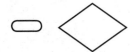 ◇

4. Count up by 2s.

0,   2,   4,   _____,   _____,   _____,

_____,   _____,   _____,   _____,   _____

LESSON 2·9 Exploring Pennies and Nickels

Write the total amount. Then show the amount using fewer coins.
Write Ⓟ for penny and Ⓝ for nickel.

(*Hint:* Exchange pennies for nickels.)

1. Ⓟ Ⓟ Ⓟ Ⓟ Ⓟ Ⓟ Ⓟ   _7_ ¢

Show this amount using fewer coins.

2. Ⓟ Ⓟ Ⓟ Ⓟ Ⓟ Ⓟ Ⓟ Ⓟ Ⓟ   _9_ ¢

Show this amount using fewer coins.

3. Ⓟ Ⓟ Ⓟ Ⓟ Ⓟ Ⓟ Ⓟ Ⓟ Ⓟ Ⓟ Ⓟ Ⓟ   _12_ ¢

Show this amount using fewer coins.

Try This

4. Ⓝ Ⓟ Ⓟ Ⓟ Ⓟ Ⓟ Ⓟ   _11_ ¢

Show this amount using fewer coins.

LESSON 2·9 **Math Boxes**

1. Record the time.

_____ o'clock

2. Use your number grid.

Start at 15.

Count up 9.

You end at _____.

Choose the best answer.

◯　6

◯　15

◯　24

◯　23

3. Circle the winning card in _Top-It_.

19　　9

4. What day of the week is today?

_____

What day of the month?

_____

What day of school?

_____

Counting Pennies and Nickels

Write the total amount.

1.

_____ ¢

2.

_____ ¢

3.

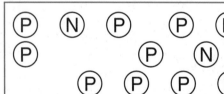

_____ ¢

Try This

4.

Write the total amount. _____ ¢

Show this amount using fewer coins.

20 twenty

Date _____

1. Use your number grid.

Start at 18.

Count back 8.

You end at _____.

2. How much money?

Ⓟ Ⓟ Ⓟ Ⓟ Ⓟ Ⓟ Ⓟ Ⓟ

Choose the best answer.

⬭ 10¢

⬭ 8¢

⬭ 16¢

⬭ 9¢

3. Draw what comes next.

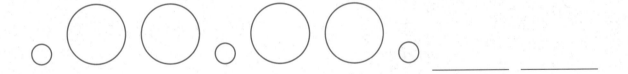

_____ _____

_____ _____

4. Count up by 2s.

2,   4,   6,   _____, _____,

_____, _____, _____, _____, _____

Date _____

1. How much money?

_____ ¢

2. Draw the hour hand.

2 o'clock

3.

The Pets We Own	
Pet	**Tallies**
Cat	~~HHH~~ ~~HHH~~
Dog	~~HHH~~ |||
Other	~~HHH~~	

How many cats? _____ cats

How many dogs?

_____ dogs

4. Count up by 10s.

20,   30,   40,

_____, _____, _____,

_____, _____, _____,

_____, _____

Date _____

1. Use your number grid.

Start at 23.

Count up 8.

You end at _____.

2. How much money?

_____ ¢

3. Draw what comes next.

_____   _____

_____   _____

4. Count up by 2s.

6,   8,   10, _____,

_____, _____, _____, _____

LESSON 2·13 School Store Mini-Poster 1

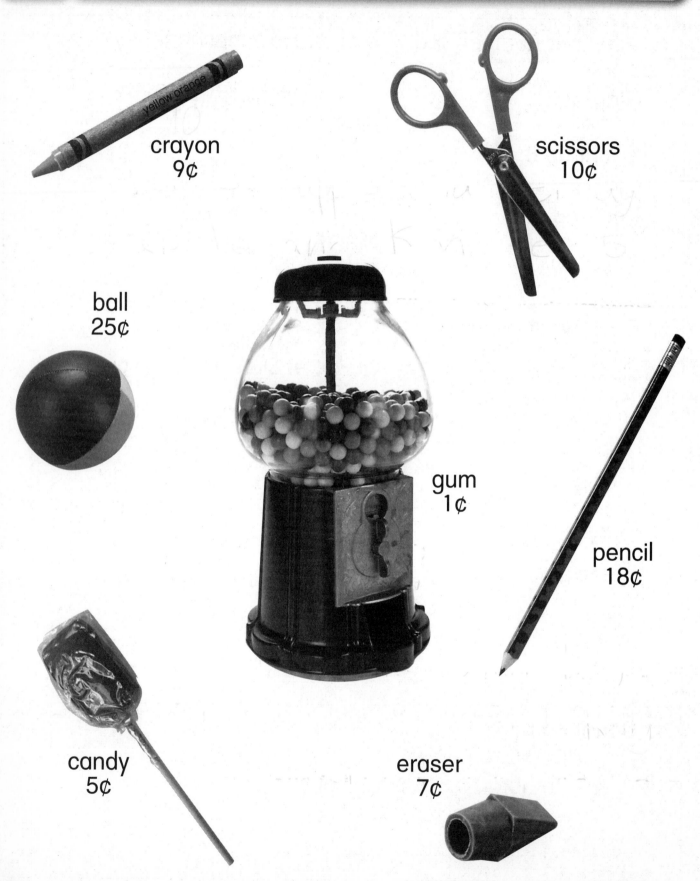

crayon
9¢

scissors
10¢

ball
25¢

gum
1¢

pencil
18¢

candy
5¢

eraser
7¢

Date _____

1. Tell how much money.

_____ ¢

_____ ¢

How much money in all? _____ ¢

2. Buy 2 items from the School Store. Draw them below.

3. Under each item you drew, show how much it costs.
Use Ⓟ for pennies and Ⓝ for nickels.

4. Circle the item that costs more.
How much more does it cost? _____ ¢

Try This

5. Draw 2 items that cost a total of 14¢.

LESSON 2·13 Creating Number Stories 1

Write a number story to go with the number sentence.
Then solve.

1. 2 + 3 + 5 = 10

Carolyn has 2 apples and Kenedy
has 3 apples and Ken hes 5
apples. Haw Many in all?

2. 4 + 6 + 8 = _____

_____

_____

3. 3 + 9 + 1 = 13   unit boox
Rings

Leighton has 3 Rings and Logan
has 9 and Emily has 1 Rings
Haw many in all?

_____

Date

LESSON 2·13

Creating Number Stories 2

Write a number story to go with the number sentence. Then solve.

1. 7 + 3 + 4 = 14 pets

My dad has 7 pets and Hannah has 3 pet and Leighton has 4. How Many in all?

2. 6 + 6 + 3 = ______

3. 9 + 2 + 4 = ______

Date _____

Math Boxes

1. Use P and N to show the cost.

28¢

Ⓝ Ⓝ Ⓝ Ⓝ Ⓝ Ⓟ Ⓟ Ⓟ

2. It is ___7___ o'clock.

Choose the best answer.

◯ 12

◯ 8

◯ 6

◯ 7

3. Fill in the table.

The Pets We Own		
Pet	Tallies	Total
Cat	⁵⁄⁄⁄ |||	8
Dog	⁵⁄⁄⁄ ⁵⁄⁄⁄ ||	12
Other	⁵⁄⁄⁄ ||||	9

4. Count back by 10s.

90,   80,   70,   60, 50, 40,

30, 20, 10

1. Write the number.

 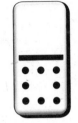

_6_   _8_

2. Count up by 10s.

30,   40,   _50_ ,

_60_ , _70_ , _80_ ,

_90_ , _100_

3. Use your template. Make a pattern with s and s.

4. Complete the number line.

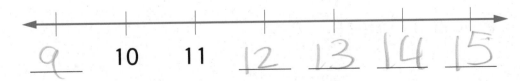

_9_   10   11   _12_   _13_   _14_   _15_

LESSON 3·1 **Patterns**

1. Draw the next 2 shapes.
Use your Pattern-Block Template.

    _____ _____

    _____ _____

     _____ _____

2. Make up your own pattern.
Then ask your partner to draw the next 2 shapes.

Try This

3. Draw the next 3 shapes.

Date _____

1. Circle the winning card in *Top-It*.

18   17

2. Draw the hour hand.

6 o'clock

3. Record the total amount.

Ⓟ Ⓟ Ⓟ Ⓟ Ⓟ Ⓟ

_____ ¢

Use Ⓟ and Ⓝ to show this amount with fewer coins.

4. What is the temperature?

Fill in the circle next to the best answer.

Ⓐ about 60°F

Ⓑ about 40°F

Ⓒ about 70°F

Ⓓ about 50°F

°F
60—
50—
40—
30—
20—
10—

LESSON 3·2 Odd and Even Patterns

How many ▢s? Label **odd** or **even**.

Example:

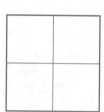

$$\underline{\quad 4 \quad}$$

$$\underline{\ even\ }$$

1.

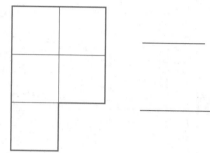

_____

_____

2.

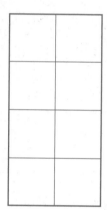

_____

_____

3.

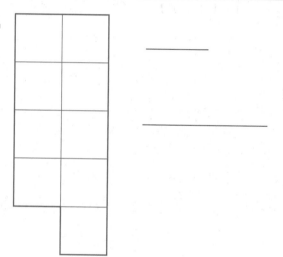

_____

_____

How many ☆s? Label **odd** or **even**.

4. ☆ ☆ ☆ ☆ ☆ ☆ ☆
☆ ☆ ☆ ☆ ☆ ☆ ☆   _____   _____

Try This

5. ☆ ☆ ☆ ☆ ☆ ☆ ☆ ☆ ☆ ☆ ☆ ☆
☆ ☆ ☆ ☆ ☆ ☆ ☆ ☆ ☆ ☆ ☆ ☆   _____

_____

LESSON 3·2 **Math Boxes**

1. Use Ⓟ and Ⓝ to show the cost.

2. What shape comes next?

Fill in the circle next to the best answer.

△ ▽ △ ▽ ____

Ⓐ ▽　　　　Ⓑ ☐

Ⓒ △　　　　Ⓓ ◯

3. Make sums of 10 pennies.

Left Hand	Right Hand
9	1
4	
	5

4. Complete this part of the number grid.

1	2			5
		13	14	
21				25
	32		34	

LESSON 3·3 The 2s Pattern

									0
1	2	3	4	5	6	7	8	9	10
11	12	13	14	15	16	17	18	19	20
21	22	23	24	25	26	27	28	29	30
31	32	33	34	35	36	37	38	39	40
41	42	43	44	45	46	47	48	49	50
51	52	53	54	55	56	57	58	59	60
61	62	63	64	65	66	67	68	69	70
71	72	73	74	75	76	77	78	79	80
81	82	83	84	85	86	87	88	89	90
91	92	93	94	95	96	97	98	99	100
101	102	103	104	105	106	107	108	109	110

Shade the 2s pattern on the above grid.

Fill in the missing numbers below.

___0___ , ___2___ , ___4___ , _____ , ___8___ , _____ ,

_____ , ___14___ , _____ , _____ , ___20___ , _____ ,

_____ , _____ , ___28___ , _____ , _____ , ___34___

Date

LESSON 3·6 Adding and Subtracting on the Number Line

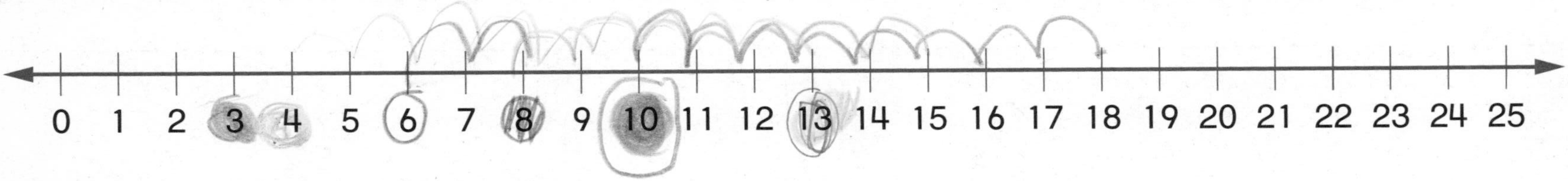

1. Start at 6. Count up 2 hops. Where do you end up?

   ______ $6 + 2 =$ 8

2. Start at 4. Count up 9 hops. Where do you end up?

   ______ $4 + 9 =$ 13

3. Start at 15. Count back 7 hops. Where do you end up?

   ______ $15 - 7 =$ 8

4. Start at 18. Count back 8 hops. Where do you end up?

   ______ $18 - 8 =$ 10

Try This

5. $5 + 8 =$ 13 6. $11 - 8 =$ 3 7. $3 + 13 =$ 16

LESSON 3·6 Math Boxes

1. Complete the table.

Before	Number	After
11	12	13
	8	
	15	
	19	

2. How many days are in a week?

Fill in the circle next to the best answer.

Ⓐ 5

Ⓑ 7

Ⓒ 10

Ⓓ 30

3. Count up by 2s.

/2,　　/4,　　/6,

——— , ——— , ——— ,

——— , ——— , ——— ,

——— , ———

4. Circle the longer one.

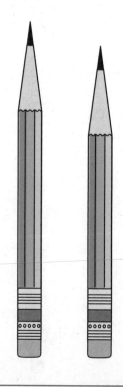

Date _____

Domino Parts and Totals

Write 3 numbers for each domino.

Example:

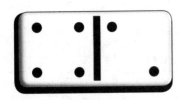

Total	
6	
Part	**Part**
4	2

1.

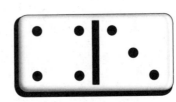

Total	
Part	**Part**

2.

Total	
Part	**Part**

3.

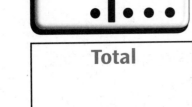

Total	
Part	**Part**

4.

Total	
Part	**Part**

5.

Total	
Part	**Part**

6. Draw dots in the domino. Write 3 numbers in the diagram.

Total	
Part	**Part**

Try This

7. Find the missing part. Draw dots in the domino.

Total	
8	
Part	**Part**
3	

LESSON 3·14   Math Boxes

1. Use your number line.
   Start at 6.
   Count back 4.

   You end at _____.

   6 − 4 = _____

2. Draw the hour hand and the minute hand.

half-past 2 o'clock

3. Write the total amount.

   Ⓓ Ⓓ Ⓝ Ⓝ Ⓝ Ⓝ Ⓟ Ⓟ

   _____ ¢

4. Write the number model.

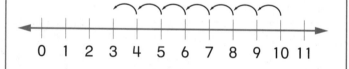

   _____

LESSON 3·15

Math Boxes

1. Color the thermometer to show about 75°F.

°F
80—
70—
60—
50—
40—

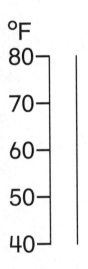

2. Circle the longer one.

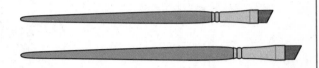

3. Complete this part of the number grid.

1	2			5
11	12			15
		23		25
			34	

4. Count up by 10s.

_50_ , _60_ ,

_70_ , _____ ,

_____ , _____ ,

_____ , _____ ,

_____ , _____

LESSON 4·1 Reading Thermometers

°F

120 —
110 —
Body Temperature → 100 —
90 —
80 —
Room Temperature → 70 —
60 —
50 —
40 —
Water Freezes → 30 —
20 —
10 —
0 —
−10 —
−20 —
−30 —
−40 —

Write the °F temperatures.

1.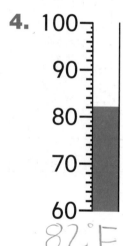

70°F

2.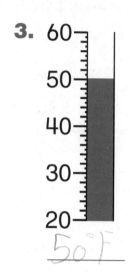

40°F

3.

50°F

4.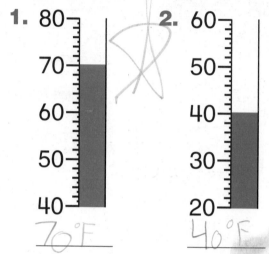

82°F

5.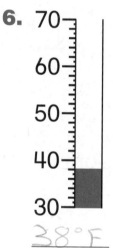

56°F

6.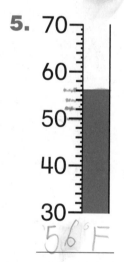

38°F

Color to show each temperature.

7. 80°F

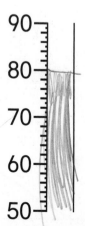

8. 62°F

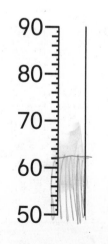

9. 58°F

Date _____

1. What is the temperature today?

_____ °F

Is the temperature odd or even?

_____

2. Write the missing numbers.

Rule
−4

| 24 | 20 | | 12 | |

3. Draw and solve.

Olga had 6 pennies.

Tyson gave her 2 more pennies.

How many pennies does Olga have now?

_____ pennies

4. Circle the winning domino in *Domino Top-It*.

LESSON 4·2 My Body and Units of Measure

Measure some objects. Record your measurements.

Unit	Picture	Object	Measurements
digit		Pincle	about 11 digits
yard		FiRe	about 2 yards
hand		Pumpkin	about 2 hands
pace		hall way	about 18 paces
cubit		block	about 2 cubits
arm span (or fathom)		Mrs. SHaft	about 2 arm spans
foot		Cube	about 5 feet
hand span		Demilola	about 14 hand spans

Date _____

My Height

Things that are taller than I am

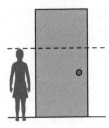

Demilola

Things that are about the same size as I am

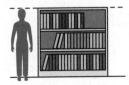

Things that are shorter than I am

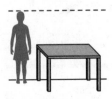

Date _____

-9	-8	-7	-6	-5	-4	-3	-2	-1	0
1	2	3	4	5	6	7	8	9	10
11	12	13	14	15	16	17	18	19	20
21	22	23	24	25	26	27	28	29	30
31	32	33	34	35	36	37	38	39	40
41	42	43	44	45	46	47	48	49	50
51	52	53	54	55	56	57	58	59	60
61	62	63	64	65	66	67	68	69	70

1. Start at 38. Count back 4. Where do you end up? __34__

$38 - 4 =$ ☐ 34

2. Start at 59. Count back 9. Where do you end up? __50__

$59 - 9 =$ ☐ 50

3. Start at 62. Count back 11. Where do you end up? __51__

$62 - 11 =$ ☐ 51

4. Start at 70. Count back 17. Where do you end up? __53__

$70 - 17 =$ ☐ 53

Try This

Subtract.

5. $43 - 20 =$ ☐ 23   **6.** $35 - 15 =$ ☐ 20

LESSON 4·2 **Math Boxes**

1. What is the temperature?
Fill in the circle next to the best answer.

○ **A.** 70°F

○ **B.** 72°F

○ **C.** 80°F

○ **D.** 76°F

°F
80
70
60
50
40

2. Draw the missing shape.

3. Complete the table.

Before	Number	After
24	25	26
	29	
	33	
	37	
	40	

4. Use a number grid.
Count by 10s.

8,   18,   _____, _____,

_____, _____, _____, _____,

_____, _____, _____

LESSON 4·3   **My Foot and the Standard Foot**

Measure two objects with the cutout of your foot.
Draw pictures of the objects or write their names.

1. I measured

It is about _____ _____ feet.
                        (your name)

2. I measured

It is about _____ _____ feet.
                        (your name)

Measure the same two objects with the foot-long foot.
Sometimes it is called the *standard foot.*

3. I measured

It is about _____ feet.

4. I measured

It is about _____ feet.

LESSON 4·3   **Math Boxes**

1. What is the temperature today?

Is the temperature odd or even?

_____ °F

_____

2. What comes next?

Rule
Count by 3s

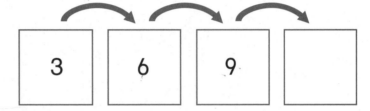

| 3 | 6 | 9 | |

Fill in the circle next to the best answer.

○ **A.** 10   ○ **B.** 11   ◉ **C.** 12   ○ **D.** 6

3. Draw and solve.

Ava had 9 pennies.

She lost 4 pennies.

How many pennies does Ava have now?

___5___ pennies

4. Circle the winning domino in *Domino Top-It.*

LESSON 4·4 Inches

Pick 4 short objects to measure. Draw or name them.
Then measure the objects with your ruler.

1.

About _____ inches long

2.

About _____ inches long

3.

About _____ inches long

4.

About _____ inches long

Date _____

1. Record the temperatures.

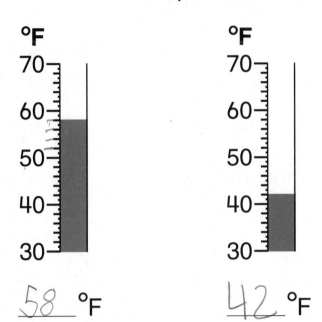

__58__ °F   __42__ °F

2. Draw the missing shape.

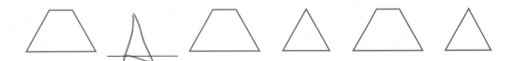

3. Complete the table.

Before	Number	After
27	28	29
34	35	36
49	50	51
100	101	102

4. Use a number grid. Count by 10s.

1, 11, 21, 31, 41, 51, 61, 71, 81, 91, 101

LESSON 4·5 **Measuring in Inches**

1. Choose two objects to measure. Estimate each object's length. Measure the objects to the nearest inch.

Object (Name it or draw it.)	My Estimate	My Measurement
	about _____ inches	about _____ inches
	about _____ inches	about _____ inches

Measure each line segment.

2. ─────────────────────

   about _____ inches

3. ───────────────────────

   about _____ inches

4. ───────────────

   about _____ inches

5. ──────────────────────────

   about _____ inches

Draw a line segment about

6. 4 inches long.

7. 2 inches long.

LESSON 4·5 **Math Boxes**

1. Measure your calculator.

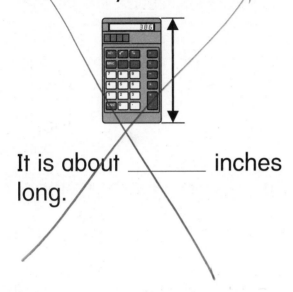

It is about _____ inches long.

2. Favorite Pets, Mr. Lee's Class

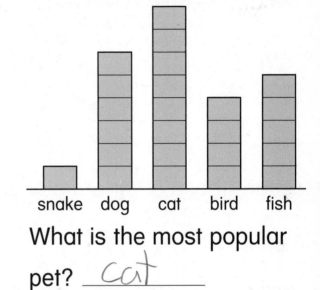

What is the most popular pet? _Cat_

How many children like snakes? _4_

3. Count back by 5s.

40, 35, 30, _25_,
_20_, _15_, _10_, _5_,
_0_

4. Write the sums.

[die 3] + [die 2] = _5_

[die 2] + [die 5] = _7_

[die 4] + [die 4] = _8_

LESSON 4·6 Measuring Parts of the Body

Record your wrist size below.

1. Wrist 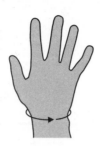 It is about _____ inches.

Measure these other parts of your body. Work with a partner.

2. Elbow  It is about _____ inches.

3. Ankle 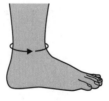 It is about _____ inches.

4. Head 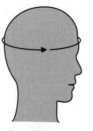 It is about _____ inches.

5. Hand span It is about _____ inches.

LESSON 4·6 Domino Parts and Totals

Find the totals.

Example:

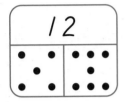

1.

2.

3.

4.

Total	
Part	**Part**
7	4

5.

Total	
Part	**Part**
2	9

6.

Total	
Part	**Part**
6	3

7.

Total	
Part	**Part**
8	8

8.

Total	
Part	**Part**
7	7

9.

Total	
Part	**Part**
8	3

Try This

Find the missing parts.

10.

Total	
12	
Part	**Part**
6	

11.

Total	
14	
Part	**Part**
	5

12.

Total	
15	
Part	**Part**
	7

LESSON 4·6 **Math Boxes**

---

1. Draw a line segment about 3 inches long.

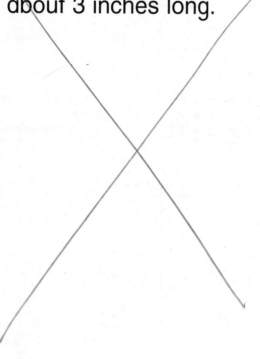

---

2. Which coins show 26¢?

Fill in the circle next to the best answer.

○ **A.** Ⓝ Ⓝ Ⓟ Ⓟ Ⓟ Ⓟ Ⓟ Ⓟ

○ **B.** Ⓓ Ⓓ Ⓟ

○ **C.** Ⓓ Ⓝ Ⓝ Ⓟ

10 5 5 1

● **D.** Ⓓ Ⓓ Ⓝ Ⓟ

10 10 5 1

20

---

3. Use your number line.

Start at 4.

Count up 5 hops.

You end at _____.

4 + 5 = _____

---

4. Tom has Ⓝ Ⓟ Ⓟ Ⓟ.

Bill has Ⓓ.

Who has more money?

_____

How much more money?

_____ ¢

Date _____

1. Today's date is _____.

My height is _____ inches.

2. This is a bar graph. It shows the heights of children in my class.

First-Grade Heights

Number of Children

7

6

5

4

3

2

1

Inches Tall

3. The "typical" height for first graders in my class is about _____ inches.

LESSON 4·7 Math Boxes

1. How long is the line segment?

_____

Fill in the circle next to the best answer.

○ **A.** about 3 inches

○ **B.** about 2 inches

○ **C.** about 1 inch

○ **D.** about 4 inches

2. Favorite Drinks, Ms. Brown's Class

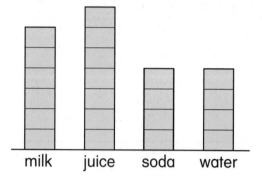

milk   juice   soda   water

How many children like milk?

_____ children

Do more children like juice or soda?

_____

3. Count back by 2s.

36,   34,   32,

____, ____, ____,

____, ____, ____,

____, ____, ____

4. Write the sums.

Date _____

Telling Time

Record the time.

1.

_____ o'clock

2.

half-past _____ o'clock

3.

quarter-past _____ o'clock

4.

quarter-to _____ o'clock

Try This

Draw the hands to show the time.

5.

half-past 3 o'clock

6.

quarter-to 5 o'clock

LESSON 4·8 Math Boxes

1. Draw a line segment about 2 inches long.

2. Show 47¢.

Use ⒟, Ⓝ, and Ⓟ.

3. Use your number line.

Start at 6.

Count up 5 hops.

You end at _____.

6 + 5 = _____

4. Nico has ⒟⒟Ⓟ.

Kenisha has ⒟ⓃⓃ.

Who has more money?

_____

How much more money?

_____ ¢

LESSON 4·9 School-Year Timeline

Think about these times of the year.
Draw pictures of things that happen during each of these months.

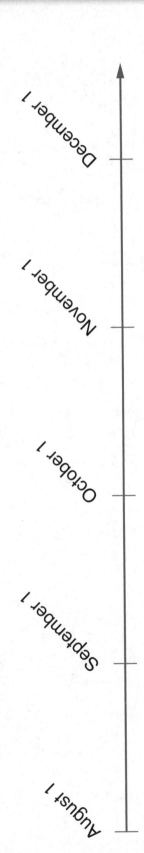

LESSON 4·9 Telling Time to the Quarter-Hour

Record the time.

1.

_____ o'clock

2.

half-past _____ o'clock

3.

quarter-after _____ o'clock

4.

quarter-after _____ o'clock

5.

quarter-to _____ o'clock

6.

quarter-to _____ o'clock

LESSON 4·9 **Math Boxes**

---

1. Measure your journal.

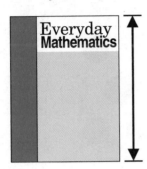

It is about _____ inches long.

---

2. What time is it?

Fill in the circle next to the best answer.

○ **A.** quarter-to 5 o'clock

○ **B.** quarter-to 4 o'clock

○ **C.** quarter-to 6 o'clock

○ **D.** quarter-to 9 o'clock

---

3. Use your number line.

Start at 8.

Count back 5 hops.

You end at _____.

$8 - 5 =$ _____

---

4. Write the sums.

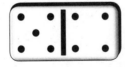

$5 + 4 =$ _____

$6 + 3 =$ _____

Math Boxes

1. Are you more likely to grab black or white?

_____

2. Write the missing numbers.

Rule

Count by 2s

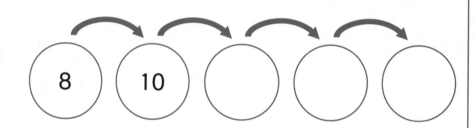

8   10   ◯   ◯   ◯

3. Record the time.

quarter-after _____ o'clock

4. Find the sums.

7 + 7 = _____

_____ = 5 + 5

Turn-Around Facts Record

1 + 6 = ___	1 + 5 = ___	1 + 4 = ___	1 + 3 = ___	1 + 2 = ___	1 + 1 = ___
2 + 6 = ___	2 + 5 = ___	2 + 4 = ___	2 + 3 = ___	2 + 2 = ___	2 + 1 = ___
3 + 6 = ___	3 + 5 = ___	3 + 4 = ___	3 + 3 = ___	3 + 2 = ___	3 + 1 = ___
4 + 6 = ___	4 + 5 = ___	4 + 4 = ___	4 + 3 = ___	4 + 2 = ___	4 + 1 = ___
5 + 6 = ___	5 + 5 = ___	5 + 4 = ___	5 + 3 = ___	5 + 2 = ___	5 + 1 = ___
6 + 6 = ___	6 + 5 = ___	6 + 4 = ___	6 + 3 = ___	6 + 2 = ___	6 + 1 = ___

LESSON 4·11 **Math Boxes**

1. Measure the line segment.

_____

It is about _____ inches long.

2. What time is it?

quarter-to _____ o'clock

3. Use your number line.

Start at 9.

Count back 7 hops.

You end at _____.

$9 - 7 =$ _____

4. Write the sums.

$1 + 5 =$ _____ $6 + 6 =$ _____

$$\begin{array}{r} 6 \\ + 4 \\ \hline \end{array}$$

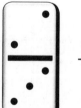

$$\begin{array}{r} 2 \\ + 3 \\ \hline \end{array}$$

Practice Making 10

1. $4 + 9 =$ _____

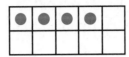

$4 + 9 =$ _____

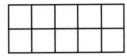

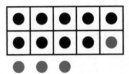

Find the sums.

2. _____ $= 8 + 3$

3. $9 + 6 =$ _____

4.
$$\begin{array}{r} 8 \\ +\,9 \\ \hline \end{array}$$

5.
$$\begin{array}{r} 8 \\ +\,7 \\ \hline \end{array}$$

6. Explain how you found the answer to Problem 5.

_____

_____

_____

Labeling True and False Number Models

Read each number model in the box. If the number model is true, write it on a line under *True*. If the number model is false, write it on a line under *False*.

$12 = 12$	$5 = 10 - 5$	$10 + 6 = 15$
$18 = 8 + 10$	$5 = 13$	$3 + 1 = 2 + 2$
$13 - 1 = 14$	$7 + 8 = 14$	$6 + 6 = 13$
	$5 + 2 = 2 + 5$	

TRUE   **FALSE**

_____   _____

_____   _____

_____   _____

_____   _____

Write your own!

TRUE   **FALSE**

_____   _____

_____   _____

LESSON 4·12 Math Boxes

1. Are you more likely to grab black or white?

_____

2. What is the rule?

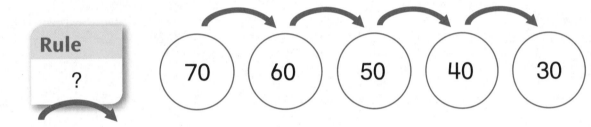

Rule

?

70 60 50 40 30

Fill in the circle next to the best answer.

○ **A.** Count up by 2s   ○ **B.** + 10

○ **C.** Subtract 5   ○ **D.** −10

3. Draw the hands.

quarter-to 6 o'clock

4. Find the sums.

$9 + 9 =$ _____

_____ $= 6 + 6$

LESSON 4·13   Math Boxes

1. Circle the winning domino in *Domino Top-It.*

2. Tim has 10¢.

Jan has 5¢.

Who has more money?

_____

How much more money?

_____ ¢

3. Write the sums.

 + = _____

_____ + _____ = _____

_____ + _____ = _____

4. Find the sums.

8 + 8 = _____

_____ = 4 + 4

LESSON 5·1 Tens-and-Ones Mat

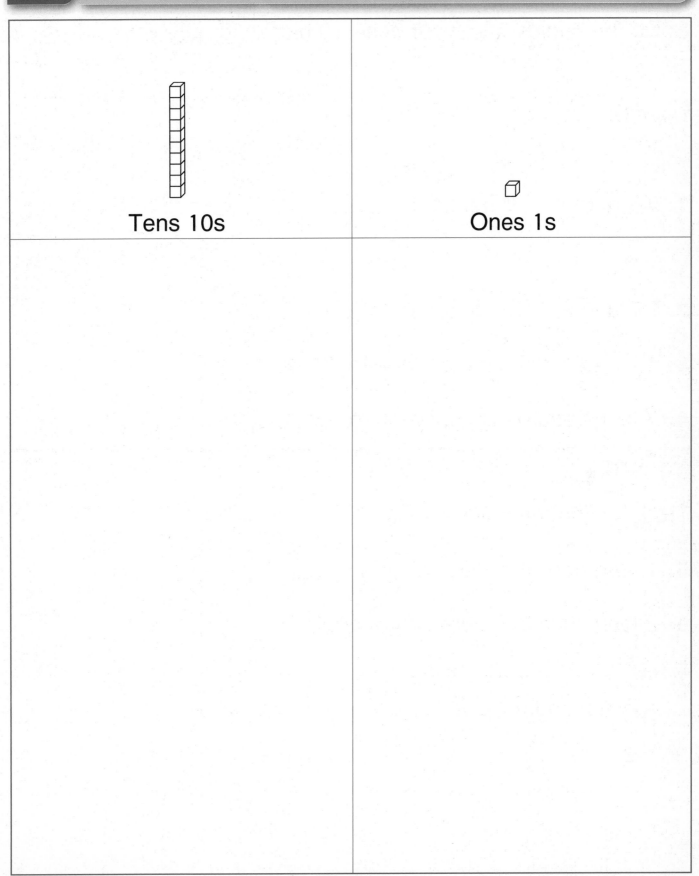

Tens 10s	Ones 1s

LESSON 5·1 — Tens-and-Ones Riddles

Solve the riddles. Use your base-10 blocks to help you.

Example: 2 ▯ and 3 ▢   What am I? __23__

1. 6 ▯ and 5 ▢   What am I? _____

2. 7 ▯ and 2 ▢   What am I? _____

3. 6 longs and 4 cubes. What am I? _____

4. 7 longs and 0 cubes. What am I? _____

Try This

Trade to find the answers.

5. 1 long and 11 cubes. What am I? _____

6. 2 longs and 14 cubes. What am I? _____

7. Make up your own riddle.
   Ask a friend to solve it.

LESSON 5·1 Math Boxes

1. Solve the riddles.

What am I? _____   What am I? _____

2. Fill in the rule and the missing numbers.

Rule
_____

(3) (6) (9) (   ) (   )

3. Add. Use a ten frame.

$8 + 3 =$ _____

_____ $= 9 + 5$

$\begin{array}{r} 7 \\ + 8 \\ \hline \end{array}$ $\begin{array}{r} 6 \\ + 9 \\ \hline \end{array}$

4. How many tallies?

卌 卌 卌 |||

_____ tallies

Odd or even?

_____

Date _____

Place-Value Mat

Hundreds

Tens

Ones

Date _____

1. Solve the riddles.

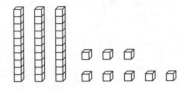

What am I? _____

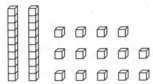

What am I? _____

2. Make a tally for 14.

_____

3. Use your number grid.

Start at 45.

Count up 13.

You end at _____.

45 + 13 = _____

4. Draw and solve.

Trey has 5 cats and 2 dogs.

How many pets does Trey have?

_____ pets

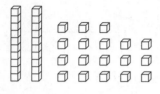

Math Boxes

LESSON 5·3

1. What am I?

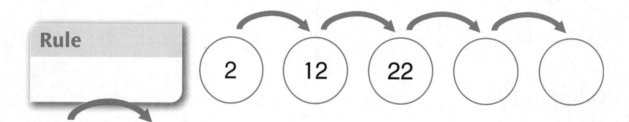

Choose the best answer.

⚬ 20   ⚬ 30

⚬ 38   ⚬ 218

---

2. Fill in the rule and the missing numbers.

Rule

(2) (12) (22) (   ) (   )

---

3. Add. Use a ten frame.

8 + 5 = _____

_____ = 9 + 3

   9   10
+ 8   + 9

---

4. Write the number.

|||| |||| |||| |||| |||| ||||

|||| |||| ||

_____

Odd or even?

_____

LESSON 5·4   **Math Boxes**

---

1. Use | and • to show the number 42.

2. How many tallies?

⊬⊬⊬ ⊬⊬⊬ ⊬⊬⊬ ⊬⊬⊬ |||

Choose the best answer.

○ 19

○ 23

○ 25

○ 43

---

3. Use your number grid.

Start at 48.

Count up 15.

You end at _____.

48 + 15 = _____

4. Draw and solve.

Rosa had 9 grapes.

She ate 3 grapes.

How many grapes does Rosa have left?

_____ grapes

LESSON 5·5 **Math Boxes**

1. Circle the tens place.

73   52   88   15   30

2. Write the number model.

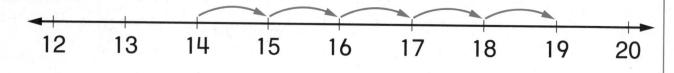

12   13   14   15   16   17   18   19   20

_____ + _____ = _____

3. Measure to the nearest inch.

_____

It is about _____ inches long.

_____

It is about _____ inches long.

4. Draw the missing dots.

Find the total number of dots.

6 + 4 = _____

7 + 3 = _____

LESSON 5·6 **"Less Than" and "More Than" Number Models**

Write < for "is less than" and > for "is more than."

1. 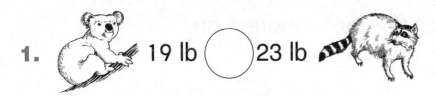 19 lb ◯ 23 lb

2. 41 lb ◯ 14 lb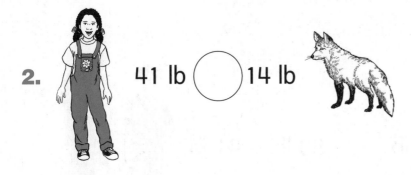

3. 75 lb ◯ 56 lb

4. 7 lb ◯ 6 lb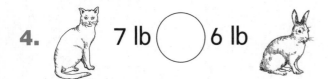

5. 50 lb ◯ 98 lb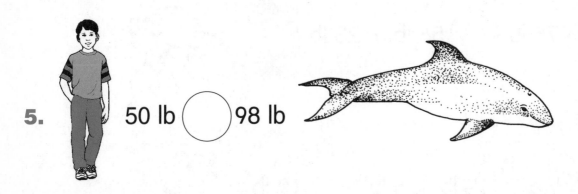

LESSON 5·6 "Less Than" and "More Than" Number Models *(cont.)*

Try This

Write < for "is less than" and > for "is more than."

6. 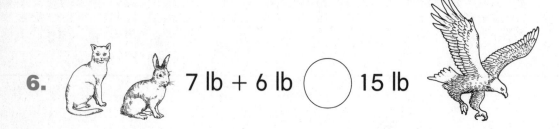 7 lb + 6 lb ◯ 15 lb

7. 120 lb ◯ 50 lb + 41 lb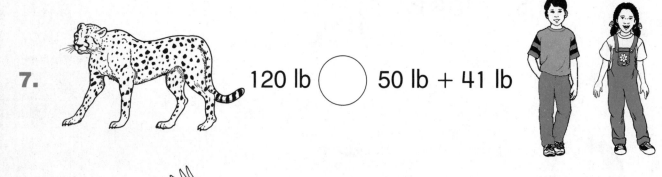

8. 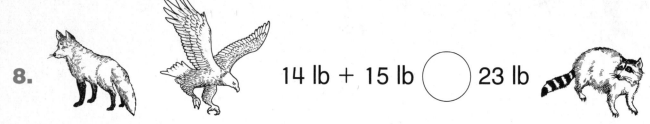 14 lb + 15 lb ◯ 23 lb

9. 75 lb ◯ 56 lb + 23 lb

10. 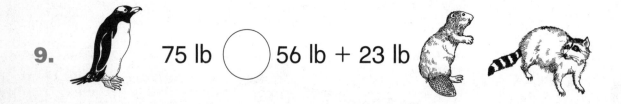 14 lb + 6 lb ◯ 19 lb

Math Boxes

1. Fill in the pattern.

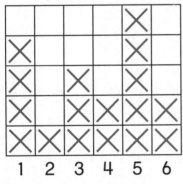

2. Show 32¢ with the fewest coins.

Use (D), (N), and (P).

3. **Judy's Dice Rolls**

1 2 3 4 5 6
Number of Dots

How many times did Judy roll a 1?

_____ times

What number did Judy roll the most?

_____

4. Draw and solve.

The garden has 4 ladybugs and 10 ants.

How many insects are there in all?

_____ insects

LESSON 5·7 How Much More? How Much Less?

Find each difference.

1. John

Nick

Who has more? ___John___ How much more? ___6___ ¢

2. June

Mia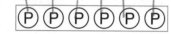

Who has less? ___Mia___ How much less? ___3___ ¢

3. Dante

Kala

Who has less? ___Dante___ How much less? ___8___ ¢

Try This

4. Carlos has 12 pennies.

Mary has 20 pennies.

Who has more? ___Mary___

How much more? ___8___ ¢

LESSON 5·7

Math Boxes

1. Circle the ones place.

12   40   6   77   54

2. Write the number model.

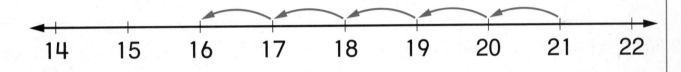

14   15   16   17   18   19   20   21   22

_____ – _____ = _____

3. How long is the line segment?

_____

Choose the best answer.

◯ about 2 inches

◯ about 3 inches

◯ about 4 inches

◯ about 5 inches

4. Draw the missing dots.

Find the total number of dots.

8 + 6 = _____

4 + 9 = _____

LESSON 5·8   Number Stories

Here is a number story Mandy made up.

I have 4 balloons.
Jamal brought 1 more.
We have 5 balloons together.

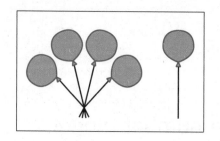

Unit
balloons

$4 + 1 = 5$

Record your own number story.
Fill in the unit box.
Write a number model.
You may want to draw a
picture for your story.

Unit

Math Boxes

1. Fill in the pattern.

 ____ ____

2. How much money?

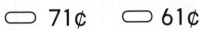

Choose the best answer.

◯ 71¢   ◯ 61¢

◯ 46¢   ◯ 40¢

3.   **Teeth Lost**

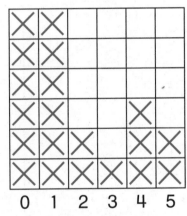

0 1 2 3 4 5
Number of Teeth

How many children have lost 0 teeth?

_____ children

How many children have lost more than 3 teeth?

_____ children

4. Draw and solve.

The chicken has 12 eggs.

2 eggs hatch.

How many eggs are left?

_____ eggs

LESSON 5·9 Dice-Throw Record

Roll a pair of dice. Draw an X in a box for the sum, from the bottom up. Which number reached the top first? _____

2	3	4	5	6	7	8	9	10	11	12

LESSON 5·9 Math Boxes

1. Write <, >, or =.

3 ☐ 13

17 ☐ 15

24 ☐ 42

28 ☐ 26

2. Draw and solve.

Meg has 8 pennies.

Maya has 3 pennies.

Who has more pennies?

_____

How many more pennies?

_____ pennies

3. How much money?

Ⓓ Ⓝ Ⓝ Ⓝ Ⓝ Ⓝ Ⓟ Ⓟ Ⓟ

_____¢

Show this amount with fewer coins.

Use Ⓟ, Ⓝ, and Ⓓ.

4. Draw the hands.

quarter-after 9 o'clock

LESSON 5·10 Using Doubles Facts

Add. Use doubles facts to help you.

1. $5 + 4 =$ _____   **2.** _____ $= 8 + 7$ **3.** $3 + 4 =$ _____

4. Explain how you solved $3 + 4$.

_____

_____

_____

5. $5 + 3 =$ _____   **6.** _____ $= 7 + 9$ **7.** $6 + 4 =$ _____

8. Explain how you solved $6 + 4$.

_____

_____

_____

1. Draw and solve.

Jade has 5 pennies.

Max has 9 pennies.

Who has fewer pennies?

_____

How many fewer pennies?

_____ fewer pennies

2. Add.

$5 + 6 =$ _____

_____ $= 7 + 8$

$$\begin{array}{r} 7 \\ + 5 \\ \hline \end{array} \qquad \begin{array}{r} 6 \\ + 8 \\ \hline \end{array}$$

3. Record the temperature.

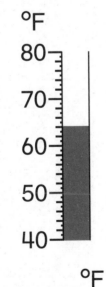

_____ °F

4. Count up by 5s.

25, _____, _____,

_____, _____, _____,

_____, _____

LESSON 5·11 Facts Table

0 +0 **0**	0 +1 **1**	0 +2 **2**	0 +3 **3**	0 +4 **4**	0 +5 **5**	0 +6 **6**	0 +7 **7**	0 +8 **8**	0 +9 **9**
1 +0 **1**	1 +1 **2**	1 +2 **3**	1 +3 **4**	1 +4 **5**	1 +5 **6**	1 +6 **7**	1 +7 **8**	1 +8 **9**	1 +9 **10**
2 +0 **2**	2 +1 **3**	2 +2 **4**	2 +3 **5**	2 +4 **6**	2 +5 **7**	2 +6 **8**	2 +7 **9**	2 +8 **10**	2 +9 **11**
3 +0 **3**	3 +1 **4**	3 +2 **5**	3 +3 **6**	3 +4 **7**	3 +5 **8**	3 +6 **9**	3 +7 **10**	3 +8 **11**	3 +9 **12**
4 +0 **4**	4 +1 **5**	4 +2 **6**	4 +3 **7**	4 +4 **8**	4 +5 **9**	4 +6 **10**	4 +7 **11**	4 +8 **12**	4 +9 **13**
5 +0 **5**	5 +1 **6**	5 +2 **7**	5 +3 **8**	5 +4 **9**	5 +5 **10**	5 +6 **11**	5 +7 **12**	5 +8 **13**	5 +9 **14**
6 +0 **6**	6 +1 **7**	6 +2 **8**	6 +3 **9**	6 +4 **10**	6 +5 **11**	6 +6 **12**	6 +7 **13**	6 +8 **14**	6 +9 **15**
7 +0 **7**	7 +1 **8**	7 +2 **9**	7 +3 **10**	7 +4 **11**	7 +5 **12**	7 +6 **13**	7 +7 **14**	7 +8 **15**	7 +9 **16**
8 +0 **8**	8 +1 **9**	8 +2 **10**	8 +3 **11**	8 +4 **12**	8 +5 **13**	8 +6 **14**	8 +7 **15**	8 +8 **16**	8 +9 **17**
9 +0 **9**	9 +1 **10**	9 +2 **11**	9 +3 **12**	9 +4 **13**	9 +5 **14**	9 +6 **15**	9 +7 **16**	9 +8 **17**	9 +9 **18**

LESSON 5·11 Easy Addition Facts

Complete.

Doubles Facts

```
   0          6
 + 0        + 6
 [   ]      [   ]

   1          7
 + 1        + 7
 [   ]      [   ]

   2          8
 + 2        + 8
 [   ]      [   ]

   3          9
 + 3        + 9
 [   ]      [   ]

   4         10
 + 4        + 10
 [   ]      [   ]

   5
 + 5
 [   ]
```

10 Sums

```
    0         [   ]
 + [   ]    +   4
   10         10

  [   ]        7
 +   9      + [   ]
   10         10

  [   ]      [   ]
 +   8      +   2
   10         10

    3        [   ]
 + [   ]    +   1
   10         10

  [   ]       10
 +   6      + [   ]
   10         10

  [   ]
 +   5
   10
```

Near Doubles Facts

Complete.

<table>
<tr><th colspan="2">Doubles-Plus-1 Facts</th></tr>
</table>

1 + 2 ☐	2 + 3 ☐
3 + 4 ☐	4 + 5 ☐
5 + 6 ☐	7 + 8 ☐
8 + 9 ☐	9 +10 ☐
6 + 7 ☐	

Doubles-Plus-2 Facts	
4 + 6 ☐	5 + 7 ☐
2 + 4 ☐	3 + 5 ☐
7 + 9 ☐	6 + 8 ☐
1 + 3 ☐	8 +10 ☐

Date

Complete.

+ 8 Facts		+ 9 Facts	
0 + 8 ☐	4 + 8 ☐	3 + 9 ☐	8 + 9 ☐
5 + 8 ☐	2 + 8 ☐	9 + 9 ☐	10 + 9 ☐
9 + 8 ☐	8 + 8 ☐	4 + 9 ☐	5 + 9 ☐
3 + 8 ☐	7 + 8 ☐	0 + 9 ☐	1 + 9 ☐
6 + 8 ☐	10 + 8 ☐	7 + 9 ☐	6 + 9 ☐
1 + 8 ☐		2 + 9 ☐	

LESSON 5·11 Math Boxes

---

1. Write <, >, or =.

   6 ☐ 8

   21 ☐ 12

   5 + 5 ☐ 10

   16 ☐ 4 + 6

---

2. Tina has Ⓝ Ⓝ Ⓟ Ⓟ Ⓟ Ⓟ.

   Fred has Ⓝ Ⓝ Ⓝ Ⓟ Ⓟ Ⓟ.

   Who has more money?

   _____

   How much more money?

   _____ ¢

---

3. How much money?

   Ⓝ Ⓓ Ⓝ Ⓟ Ⓓ Ⓝ Ⓝ

   _____ ¢

   Show this amount with
   fewer coins. Use Ⓟ, Ⓝ,
   and Ⓓ.

---

4. Draw the hands.

   quarter-past 8 o'clock

LESSON 5·12 "What's My Rule?"

Find the rules and missing numbers.

1. in ↓

Rule

out ↓

in		out
7	→	4
11	→	8
4	→	1
9	→	___

Your turn: ___ → ___

2. in ↓

Rule

out

in		out
5	→	10
8	→	13
12	→	17
16	→	___

Your turn: ___ → ___

3. in ↓

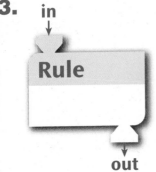

Rule

out ↓

in		out
33	→	43
12	→	22
27	→	37
9	→	___
24	→	___

Your turn: ___ → ___

Try This

4. in ↓

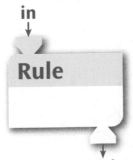

Rule

out

in		out
1	→	2
2	→	4
3	→	6
4	→	___
6	→	___

Your turn: ___ → ___

LESSON 5·12 Math Boxes

1. Ray has Ⓓ Ⓝ Ⓝ Ⓟ Ⓟ.
Dee has Ⓓ Ⓓ Ⓟ Ⓟ Ⓟ Ⓟ.

Who has more money?

_____

How much more money?

_____ ¢

2. Add.

$6 + 7 =$ _____

_____ $= 4 + 5$

$\begin{array}{r} 7 \\ + 5 \\ \hline \end{array}$ $\qquad$ $\begin{array}{r} 9 \\ + 7 \\ \hline \end{array}$

3. What is the temperature?

°F

Choose the best answer.

◯ 60°F   ◯ 72°F

◯ 68°F   ◯ 74°F

4. Count back by 5s.

45, _____, _____,

_____, _____, _____,

_____, _____

Date _____

Find the rule.

1. in ↓ Rule → out

in	out
3	5
12	14
10	12

Your turn: _____

2. in ↓ Rule → out

in	out
4	1
12	9
17	14

Your turn: _____

3. What comes out?

in ↓ Rule +10 → out

in	out
3	13
16	
25	

Your turn: _____

4. Make your own.

in ↓ Rule → out

in	out

LESSON 5·13 Math Boxes

---

1. Write <, >, or =.

28 ☐ 38

34 ☐ 43

6 + 7 ☐ 12

16 ☐ 10 + 6

---

2. Lois has Ⓟ Ⓟ Ⓓ Ⓝ Ⓓ Ⓝ.
Joe has Ⓓ Ⓟ Ⓟ Ⓟ Ⓟ Ⓝ.

Who has more money?

_____

How much more money?

_____¢

---

3. How much money?

Ⓓ Ⓝ Ⓓ Ⓟ Ⓟ Ⓟ Ⓟ Ⓟ Ⓝ

_____¢

Show this amount with
fewer coins. Use Ⓟ, Ⓝ,
and Ⓓ.

---

4. It is quarter-to _____.

Choose the best answer.

⬭ 9 o'clock   ⬭ 1 o'clock

⬭ 10 o'clock   ⬭ 12 o'clock

Date _____

1. Draw and solve.

Yuko has 3 red balloons, 4 green balloons, and 1 blue balloon.

How many balloons does she have in all?

_____ balloons

2. Draw the hands.

quarter-after 7 o'clock

3. Draw the missing dots.

Find the total number of dots.

5 + 7 = _____

8 + 3 = _____

4. Count up by 5s.

5,   10,   _____,

_____, _____, _____,

_____, _____, _____

Number Cards 0–15

15	**14**	**13**	**12**
11	**10**	**9**	**8**
7	**6**	**5**	**4**
3	**2**	**1**	**0**

Number Cards 16–22

16	**17**	**18**	**19**
20	**21**	**22**	**+**
—	**×**	**÷**	**=**
<	**?**	**wild card**	**wild card**

Number Cards 0–9

1 0

5 4 3 2

9 8 7 6

Clock Face, Hour and Minute Hands

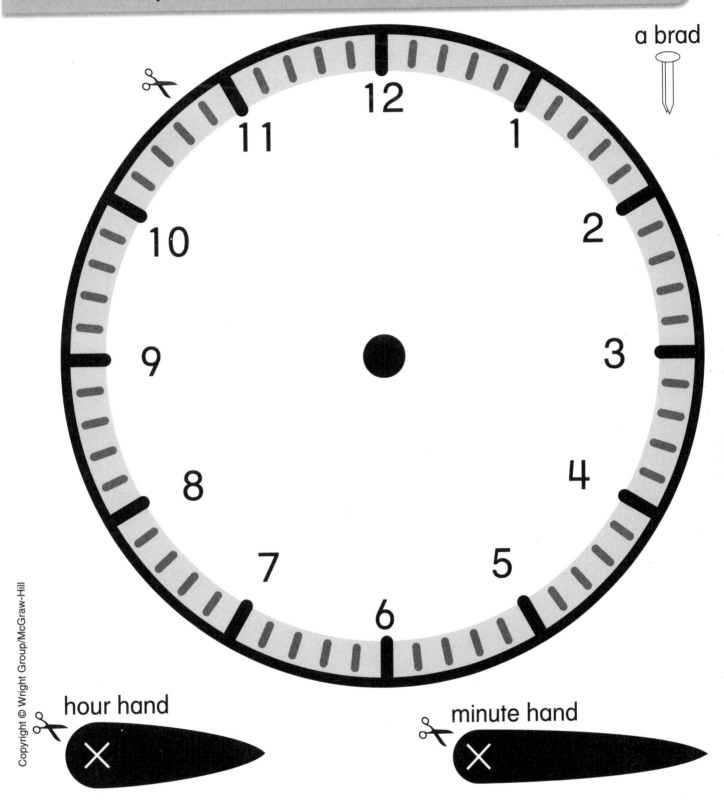

a brad

hour hand

minute hand

Activity Sheet 3

Domino Cutouts

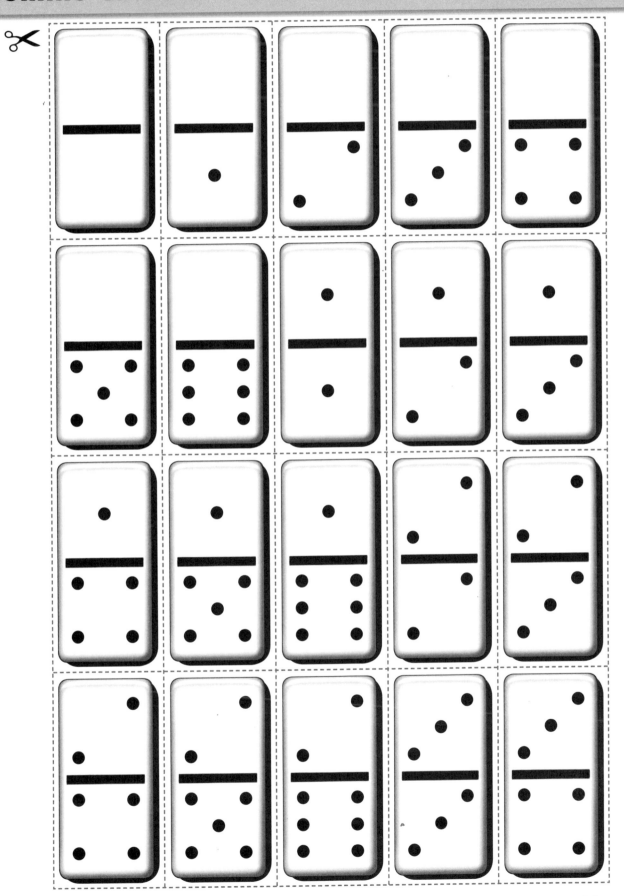

Domino Cutouts

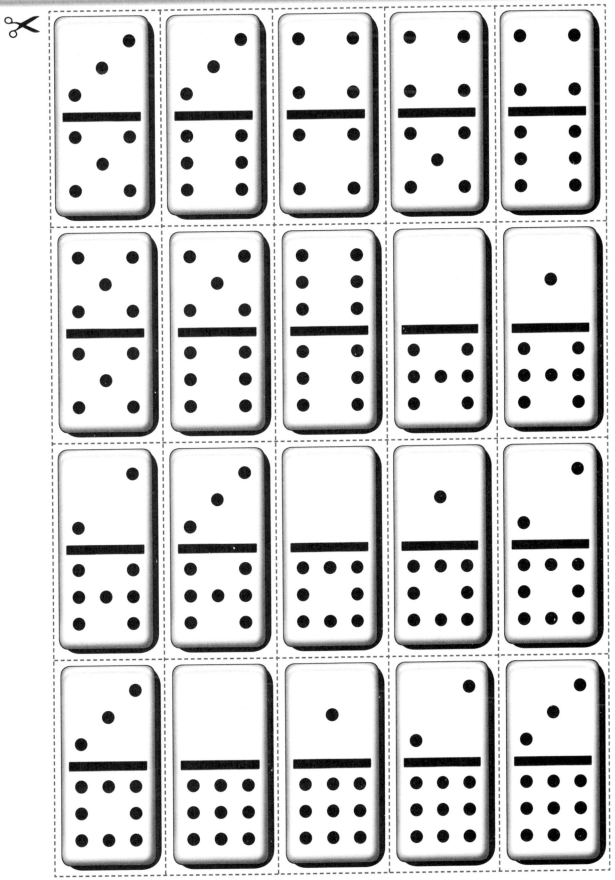

Place-Value Mat

Hundreds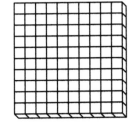

Tens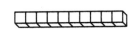

Ones ◻

Animal Cards

First-grade girl
41 lb

7-year-old-boy
50 lb

Cheetah
120 lb

Porpoise
98 lb

Penguin
75 lb

Beaver
56 lb

Animal Cards

7-year-old boy
50 in.

First-grade girl
43 in.

Porpoise
72 in.

Cheetah
48 in.

Beaver
30 in.

Penguin
36 in.

Activity Sheet 7

Animal Cards

Cat
7 lb

Fox
14 lb

Koala
19 lb

Raccoon
23 lb

Rabbit
6 lb

Eagle
15 lb

Animal Cards

Fox
20 in.

Cat
12 in.

Raccoon
23 in.

Koala
24 in.

Eagle
35 in.

Rabbit
11 in.

Activity Sheet 8